DON'T PUT GOD IN A BOX

Reflections on God and the Universe

Frank Flynn

authorHOUSE

1663 LIBERTY DRIVE, SUITE 200
BLOOMINGTON, INDIANA 47403
(800) 839-8640
WWW.AUTHORHOUSE.COM

© 2005 Frank Flynn. All Rights Reserved.

No part of this book may be reproduced, stored in a retrieval system, or transmitted by any means without the written permission of the author.

First published by AuthorHouse 11/16/05

ISBN: 1-4208-8681-9 (sc)

Printed in the United States of America
Bloomington, Indiana

This book is printed on acid-free paper.

All Bible references are from the 'New International Version' unless otherwise stated.

Acknowledgements

I would like to thank my wife Sue and my three sons, Mike, Nick and Andrew for being helpfully critical of my efforts to write a book. Also my good friends and mentors Rev. Mike Ranyard and Rev. John Elliston for keeping a watchful eye on my theological ideas and correcting me when I strayed, John Pearce for his thoughtful comments and Malcolm Dale for his tactful but much needed advice on my English, and painstaking scrutiny of the final text. All these contributions have been very much appreciated.

Photographs are reproduced by courtesy of David Malin Images (Anglo-Australian Observatory, Akira Fujii and David Miller), and NASA

a spiral galaxy	DMI (Anglo-Australian Observatory)
the Great Nebula in Orion	DMI (Anglo-Australian Observatory)
the Crab Nebula	DMI (RGO/IAC)
the Sombrero Galaxy	DMI (Anglo-Australian Observatory)
a field of distant galaxies	NASA (Hubble space telescope)
sunrise glow	DMI (David Miller collection)
Mars	NASA (Hubble space telescope)
transit of Venus	author's telescope (photo John Pearce)
constellation of Orion	DMI (Akira Fujii collection)
asteroid Eros	NASA (NEAT spacecraft)
Earth from space	NASA (Apollo 17 mission)
Aldrin on the Moon	NASA (Apollo 11 mission)
comet Hale-Bopp	DMI (Akira Fujii collection)
total eclipse	DMI (Akira Fujii collection)
Anglo-Australian telescope	DMI (Anglo-Australian Observatory)

Table of Contents

Acknowledgements v

Introduction   Setting the Scene. xi

Theme 1   Beginnings and Endings 1

Theme 2   Let there be Light 11

Theme 3   God of the Infinite 17

Theme 4   Our Protective Blanket 23

Theme 5   The Question of Life 27

Theme 6   Are We Alone? 31

Theme 7   Man's Conceit and Impotence 39

Theme 8   Send a Fireball 47

Theme 9   Small is Beautiful 53

Theme 10   Man in Space 57

Theme 11   Star of Bethlehem 63

Theme 12   Total Eclipse 69

Conclusion   Drawing the ends together 75

Introduction   Setting the Scene

*The beautiful spiral galaxy NGC 2997 in Antlia.
Light takes hundreds of millions of years to reach us from this
object*

> *When I consider your heavens, the work of your fingers, the Moon and the stars, which you have set in place, what is man that you are mindful of him, the son of man that you care for him? You made him a little lower than the heavenly beings and crowned him with glory and honour.*
>
> Psalm 8:3–5

For a long time I have wanted to write these thoughts about God, and the universe he has created. This is not a scientific textbook, and even less a sort of theological treatise. It is simply a set of personal reflections which for years have been spinning around in my head. Now at last, in retirement, I have the chance to get it all down in writing. So here goes.

I had a scientific education and have spent much of my life as a teacher of astronomy. I am also a committed Christian. As I get older I find myself becoming more and more frustrated on the one hand with those Christians who persistently duck scientific issues, maybe because they feel a little overawed by them, or perhaps because they are worried they are heading for a conflict of belief. On the other hand there are many scientists, who rather arrogantly assert that science is incompatible with a belief in God.

Einstein wrote:

> *"Science without religion is lame, religion without science is blind."*

I really believe science and religious faith can progress happily side by side. They both demand total honesty, and are pursuing different but parallel agendas. The Bible tells us about God's relationship with the human race, which he created. It is emphatically not a scientific text book, and should not be treated as such. Scientists in pursuing their observations and theories about the natural universe also have to remember they are studying the created order, and not the Creator himself.

Galileo, father of modern experimental science and himself a devout Catholic said

> *"the bible was intended to teach us*
> *how to go to heaven, not how heaven goes".*

So this is intended rather to be a short devotional work, bringing together in harmony some reflections about the Creator, and the wonders of his creation. This is not to say all issues are neatly wrapped up and easy to explain. All Christians grapple with mysteries they do not understand, and every scientist would agree that as soon as one scientific problem is solved two more jump up to take its place!

The book is divided into short themes. These are essentially free-standing, although there is a rough underlying progression from man's increasing awareness of God and the vastness of his universe, to his contemplation of his own smallness and utter helplessness in the universe, and then the amazing reality of God himself actually caring about us, and communicating with us through the person of Jesus.

Each theme starts with some astronomical background, and ends with a short personal reflection.

Theme 1 Beginnings and Endings

The Great Nebula in Orion (M42) A region of star formation

Frank Flynn

*The Crab nebula, a supernova remnant.
An old massive star which exploded at the end if its life cycle*

In the beginning God created the Heavens and the Earth

Genesis1:1

The subject of cosmology[1] is very young. Even at the end of the 19th century the wisdom of the day was that our own galaxy of stars[2] the Milky Way[3] was the sum total of the universe. As bigger and better telescopes revealed strange spiral shaped **nebulae**[4] it was assumed at first these had to be giant gas clouds embedded in the Milky Way itself. Only with advances in spectroscopy [5] astronomers were enabled to actually measure the speeds and distances of these objects. Then it was realised the spirals were actually distant galaxies in their own right. Many resembled the Milky Way Galaxy itself, but were often much larger, and at mind-blowing distances. So only a little over a hundred years ago our concept of the scale of the universe took a gigantic leap forward, and modern cosmology began.

To set the scene let us take a unit of distance which will help us get more easily around the universe. Light travels so fast we think of it as almost instantaneous, but its speed, of approximately 300 000 km/sec, has been determined with great precision. This means light takes just over 1 second to reach us from the Moon. Distances in the solar system[6] are a matter of minutes or hours of light-time, for example 8 minutes from the Sun, 40 minutes from Jupiter and about 5 hours from Pluto. But as we go further afield we find distances are mind-blowing. Within the Milky Way Galaxy light takes 4 years to reach us from even the nearest star, Proxima Centauri, 30,000 years from the centre of the galaxy and 100,000 years to merely cross the diameter of our galaxy from one extremity to the other - more than the span of human civilization – compare this with just a second from the Moon!

Even nearby galaxies beyond our own in the so called local group, our own galactic family, have light times of up to several million years. When we venture beyond this, to galaxies in more remote clusters beyond the comfort zone of the local group light times are measured in hundreds, and then thousands of millions of years.

During the 20th century cosmologists began to advance theories to explain the structure and evolution of the universe. One key observational fact which all astronomers subscribe to is that the universe is expanding. This is based on very good spectroscopic evidence that the galaxies are mutually receding. Whatever galaxy you are on, all the others are moving away from you. This is the famous **red-shift** principle. The lines in the spectrum of light from the galaxy are slightly shifted in position towards the red end of the spectrum, that is they are increased in wavelength. The Doppler Principle tells us this means the object is receding. Moreover the speed of recession increases as the distance of the galaxies increases, according to a law established by the famous cosmologist Hubble[7] in 1929.

A helpful analogy in understanding the recession of the galaxies is to consider dots painted on a balloon. As the balloon expands each dot moves away from every other dot. This illustrates that the recession is not just outwards from any one special galaxy, but a **mutual** recession of every galaxy from every other one. Early last century Sir Arthur Eddington, advanced the Evolutionary or 'Big Bang' theory of the universe. According to this view, the universe was created with a 'big bang' and the debris of the explosion of creation has been receding further and further out into space over the thousands of millions of years which have ensued since then.

In an exciting and controversial period during the 1960s Fred Hoyle and others advanced a completely different concept of the universe - the Steady State theory. This suggested that instead of starting with a 'big bang' the universe had always been there, and always will be. To compensate for the thinning out of the universe as the galaxies recede, 'continuous creation' was taking place to restore the balance. This notable theory was finally discarded due to the discovery of **cosmic background radiation**[8]. The universe was found by radio astronomers to have a steady background temperature of about 3 degrees above absolute zero[9]. This corresponds to the predicted residual

Don't Put God In A Box

temperature left over from the Big Bang itself. Just as a kettle is still slightly warm a long time after it has been switched off, so the universe, even after thousands of millions of years still has a warm 'glow'. Actually 3 degrees above absolute zero, that is about -270 degrees Celsius, isn't exactly warm, but nevertheless it is a little above absolute zero!

So now in the early 21st century cosmologists are still refining various models to explain the universe as we see it. All are firmly based on the broad idea of the Big Bang theory, but with many subtle variations.

In the cosmos there is a conflict between two physical forces - the blast of the initial explosion of the 'big bang', causing the universe to expand and the galaxies to recede from each other, and the force of gravity. Gravity, which depends on mass, is the most important physical force in the universe. It acts inwards, attempting to oppose expansion and hold the universe together. It is biding its time, waiting for the blast of creation to subside, so that it can begin to contract the universe together again. Will this happen? We don't know. It all depends on the amount of mass in the universe. If the total amount is more than a certain critical value, then gravity will have its way, put a break on the expansion process and cause the universe to end up with a 'big crunch'. If on the other hand the amount of mass is less than this critical value, gravity will not be strong enough to pull the galaxies together again and the universe will become static, or go on gently expanding for ever.

Within this cosmic framework stars too have an evolutionary cycle all of their own. They form from condensation under gravity of a huge gas cloud – a **diffuse nebula**[10] and have a long stable period of 'middle age' while hydrogen is being gradually converted into helium to produce energy, and then eventually, when most of the star's energy has been used up, the star gradually begins its 'old age' and disintegrates.

The end-state of a star can take several forms. A very massive star can have a violent end in a **supernova**[11] explosion, but a star with a more modest mass like the Sun will expand more gently to many times its present diameter to become a

red giant[12]. It seems certain that this will happen to the Sun one day – but in thousands of millions of years time! When it does happen however it will have serious implications for the inner planets including the Earth. The Sun will swell many times in size, engulf the orbits of Mercury and Venus, and become very uncomfortably close to the Earth itself. The decaying Sun will then throw off a lot of its matter, which will dissipate into space in the form of rings of gas, rather like smoke rings puffed from a cigarette, forming a ***planetary nebula.***[13] Eventually the core of the Sun will reach its final death-state as a very dense small dark 'husk' of its former self, a **white dwarf.**[14]

*****

It is quite impossible that life on Earth and indeed the Earth itself could survive the death-throes of the Sun. Life on Earth is absolutely dependent on the Sun's energy and so we cannot isolate our thinking about life, and the destiny of mankind, from a study of the universe.

Science and human life are inextricably linked, but many questions arise about how the theories of cosmology impact on the Christian faith – what about creation, for example, and what about the 'end times'?

We read in Isaiah 60:19-20

> *The Sun will no more be your light by day, nor will the brightness of the Moon shine on you, for the Lord will be your everlasting light, and your God will be your glory. Your Sun will never set again, and your Moon will wane no more; the Lord will be your everlasting light, and your days of sorrow will end.*

and in Revelation 21:23

The city does not need the Sun or the Moon to shine on it, for the glory of God gives it light, and the Lamb is its lamp.

Some scientists are quite happy to accept the comfortable enough concept of a 'creator God', but cast him simply in the role of 'the unknowable remote initiator' – the one who 'winds up the spring and lets it go'. They find it difficult to conceive that he could continue to be in any way involved in his creation in an ongoing way. This view, although at least 'deistic', completely negates the Christian understanding of a God who is still intimately involved in the universe he created.

The Bible is totally devoted to this relationship between God and his creation, and especially between God and humankind. The Jews have plenty of evidence of God's ongoing engagement with mankind, from the Tenach, the Old Testament, and Christians have in addition the New Testament, and the model of Jesus.

The disciple Philip said:

"Lord, show us the Father, and that will be enough for us",

and Jesus answered:

"Don't you know me, Philip, even after I have been among you such a long time? Anyone who has seen me has seen the Father"

John 14:8-9

When human beings create something – a work of art, a piece of music, or a fine building – it remains a once-and-for-all creation. It does not change its original form. It is true it could later be modified in some way, but a human creation is essentially static. The painting or the music does not have a will of its own. But Christians believe God created Life, which is dynamic and something altogether on a higher plane. And God created Human Life, capable of having a relationship with the Creator himself.

The issue of life, and the ongoing search for evidence of life elsewhere in the universe, are considered in themes 5 and 6.

Have you ever been on the beach, watching the breakers roll in from the sea, and wondered when they began, and when if ever they will stop? Have you felt that your own life-span, and indeed the life span of the whole of mankind, is just a flash, and that in millions of years time, when we are long forgotten, and when possibly humankind no longer inhabits this planet, these breakers will still be rolling in?

God the Creator is the one permanent living presence, and Heaven transcends the physical universe. We should not imagine our little planet, or indeed our parent Sun, will last forever. Christians have traditionally expected a final judgement of all people and the consequent end of this creation. But it may be wrong to assume that at the 'end time' of human life the physical universe itself will necessarily cease to exist. The physical universe may perhaps continue to be, and the waves continue to crash on the beaches. Maybe God has indeed made an infinite universe, which will just go on and on. Who are we to dictate the times and the seasons?

*****

1   cosmology – the study of the universe as a whole.
2   galaxy – a major grouping of often many millions of stars.
3   Milky Way – the name of our own galaxy – often seen as a band of hazy whiteness straddling the sky on a clear night.
4   nebula – an indistinct 'fuzzy' object as seen in the telescope, usually a gigantic gas cloud.
5   spectroscopy – the study of light split into its component colours, each with its own particular wavelength.
6   solar system – the system of objects (planets, asteroids, comets and meteors) which are attracted by the Sun's gravity and travel around it in orbits.

7   Hubble's Law – the proportionality between the distance of a galaxy away from us, and its speed of recession ('the further the faster').
8   cosmic background radiation – the universe has a background 'warm glow' with a temperature of 3 degrees above absolute zero (i.e. 3deg Kelvin, which equals - 270 deg Celsius).
9   absolute zero – the lowest temperature that can be (-273 degrees Celsius).
10 diffuse nebula – a gigantic cloud of hydrogen, out of which stars are formed by gravitational attraction.
11 supernova - a very massive star which has blown up at the end of its life cycle and formed a very bright nebula of gas from the debris of the explosion.
12 red giant – a very large, cool red star, nearing the end of its life.
13 planetary nebula – rings of gas thrown off a decaying star showing up as bright halos around the original star.
14 white dwarf – the remnants of a star which has expended all its energy. It is very small and dense, and only faintly glows.

Theme 2   Let there be Light

The Sombrero Galaxy (M104), in the Virgo Cluster of Galaxies
An intense source of light, radio waves and other radiation

Frank Flynn

> *And God said "Let there be light", and there was light. God saw that the light was good, and he separated the light from the darkness.*
> *God called the light 'day', and the darkness he called 'night'.*
>
> *Genesis 1:3 - 5*

What we understand as 'light' is just one small band in the spectrum of electromagnetic radiation[1] which permeates the universe. We recognise light, or radio waves, infra-red, ultra-violet, X-rays or gamma rays according to the wavelength of the emission. Radio waves are at the long wave end of the spectrum, gamma rays at the short end and light somewhere in the middle. All types of electromagnetic radiation travel at the same speed – the speed of light – approximately 300,000 km per sec.

As our eyes are adapted to picking up light it is not surprising we regard light as the principal illuminator of the universe. But only comparatively recently astronomy has opened itself out to studying the universe in wave bands other than visible light. Radio Astronomy came into being during the second half of the last century. Light and radio waves can penetrate the atmosphere so that observations can be made from ground-based telescopes, but it is only since we have been able to get above the atmosphere and make observations from spacecraft that it has been possible to study the universe in other wavebands too.

Have you ever considered why the sky is dark at night? This sounds like a silly question until you come to consider that there is an immense number of stars and galaxies up there all radiating light. Thousands of millions of bright stars and galaxies are shining down on us, and although due to enormous distances many of them are simply small pin points, this is balanced out by the sheer number of them. Images from the Hubble Space Telescope show that there is virtually a galaxy at the end of any random line of sight out into space.

This famous paradox, known as Olber's Paradox was first advanced in 1826, and puzzled astronomers for many years, until it was realised the idea of the 'expanding universe' held the key. As the galaxies recede their light is thinned out and diminished in intensity, so the contributions of the more distant galaxies are not sufficient to swamp the sky with light. If the universe were contracting, or even static, this would be a very different story and we would not be able to withstand the blinding brightness of the sky.

Cosmologists are now also very interested in 'dark matter'[2]. With hindsight it is astonishing that until quite recently astronomers assumed the totality of the mass of the universe was made up of matter which could only be detected because it emitted radiation in the form of light or radio waves.

Dark matter does not radiate energy and so cannot be detected by conventional observation. But dark matter does give its presence away indirectly. If its mass is significant it will have a gravitational effect on a passing light beam causing it to bend slightly, and this effect can be observed.

The study of dark matter is immensely important to cosmologists. They can use this to estimate the proportion of dark matter to the total mass of the universe. It gives them some indication of where we stand in the 'power struggle' between the forces of expansion (the blast of the Big Bang) on the one hand, and gravity on the other. The more mass the greater the overall power of gravity to contract the universe again to an ultimate 'crunch'.

*****

God said "Let there be lights in the expanse of the sky to separate the day from the night, and let them serve as signs to mark seasons and days and years, and let them be lights in the expanse of the sky to give light on the Earth". And it was so.

Frank Flynn

> *God made two great lights – the greater light to govern the day and the lesser light to govern the night. He also made the stars.*
> *God set them in the expanse of the sky to give light on the Earth, to govern the day and the night, and to separate light from darkness.*
> *And God saw that it was good.*
> *Genesis 1:14-18*

We take light for granted, but light is the result of energy generation, and without this process the universe remains dark.

God has given us light to illuminate our world, but we are thoughtless and squander our light, and other forms of energy, causing massive pollution. Satellite photographs of landmasses at night show the extent of our light-polluted planet. The sea shores of Europe stand out like a sharply drawn map and the large conurbations give a continuous fuzzy brightness. In London the night sky glows like daytime and city children never see a starry sky.

Imagine how it would be in a dark world. It is bad enough when our power supply fails and we are left to grope around looking for the candles. We are used to living in an illuminated world. All our immediate light comes directly or indirectly from the Sun. So does our heat and all other forms of energy. In a very real sense the Sun is our 'parent' body.

We may have some sympathy for the animists of ancient times, who worshipped the Sun. They rightly understood their life-line in a physical sense, for light and heat and all other forms of energy needed by earthlings, had its origin in the Sun. We are connected to the Sun with an umbilical cord without which no life form could survive on the Earth. But it is sad that, like the animists, we ourselves do not always recognise the living Father God who created the process which gave birth to our Sun and so gave us light. And life.

*****

1   electromagnetic radiation – energy which passes through a vacuum, such as light, radio waves, or X rays. All these types of radiation have a common speed (300,000 km. per sec), and differ only in their wavelength.
2   dark matter – matter which does not emit electromagnetic radiation and so cannot be detected by conventional means. It can however be detected indirectly as its mass has a gravitational effect on light, and slightly bends its path.

Theme 3   God of the Infinite

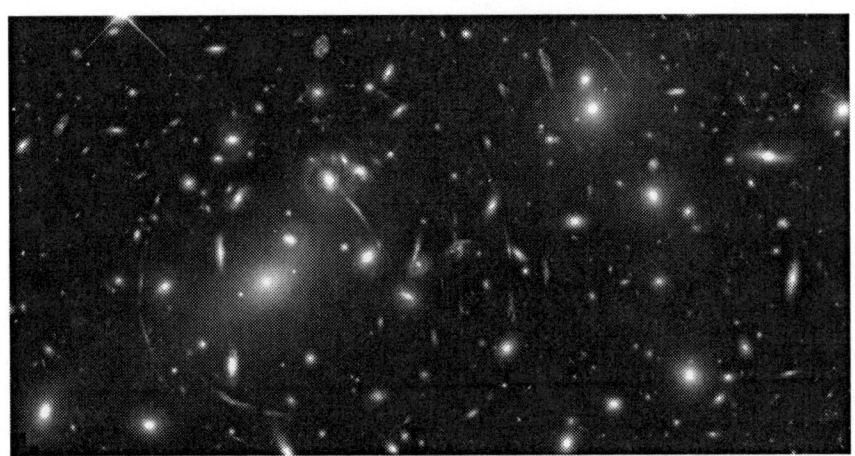

A field of remote galaxies seen with the Hubble Space Telescope. Light can take over a thousand million years to reach us from remote sources such as these

Frank Flynn

> *As far as the east is from the west...*
> *Psalm 103:12*

Is the universe finite or infinite? This is a question which has intrigued cosmologists for many years. On the face of it an expanding universe following an initial Big Bang would imply a finite universe, whereas the steady state model referred to in the first theme requires it to be infinite.

During the last hundred years more and more remote galaxies have been discovered, with light times in excess of ten thousand million years. Every direction line out into space seems to point to a remote galaxy. The universe does indeed seem to be limitless as far as our own concept of distance is concerned.

There is actually a sort of half-way position between finite and infinite. This is the concept of a universe which is 'finite but unbounded'. It all depends on the number of dimensions we can conceive in our minds. Imagine for example a tiny spider walking on the surface of a perfectly smooth ball. Suppose this little creature can only think in two dimensions. Then his world seems to be flat and as he walks ceaselessly round and round this little ball he thinks he is in infinite flatland. There is no end to his travels. We superior beings can think in three dimensions so we laugh at the poor little spider as we know his universe is really finite – a mere little globe.

So it may be with our universe – boundless in three dimensions, which is all we can conceive physically, but does this really mean it is infinite?

A useful device, in trying to get the whole universe onto one sheet of paper, is to employ a **logarithmic scale**[1]. This is to label each line of your page on a scale which multiplies each previous distance by 10, so that instead of labelling your lines to represent distances on a conventional scale 1,2,3,4,5, you will label them successively 1, 10, 100, 1000, 10000. In this way each new band of distance is compressed and you can represent on the same sheet nearby objects such as the Moon and the planets of the solar system at the top of the

Don't Put God In A Box

page, together with objects in our own the galaxy the Milky Way further down, and then even very remote galaxies towards the bottom of the page. This enables us to cope with the immensity of these extragalactic[2] distances. It really tells us nothing about where it ends. But it looks as if a natural limit does occur when the speed of recession of a galaxy is so great it approaches the speed of light.

This logarithmic concept can be made to work in both directions. You can extend your page **upwards** if you wish to chart successively smaller and smaller distances, 1/10, 1/100, 1/1000, until you get to electrons, and all the other fundamental particles. So using a logarithmic scale actually places us somewhere in the middle, rather than very much at the 'bottom end' of the distance scale, where we normally think we are.

*****

> *"But do not forget this one thing, dear friends: With the Lord a day is like a thousand years, and a thousand years are like a day"*
> 2 Peter 3:8

I find it very helpful to remember that our time and distance scales are strictly relative to **ourselves.** We probably think a butterfly has a very short life, but maybe the butterfly, within her own frame of reference, considers her life long and packed with fulfilment! Consider also the huge 'jungle' of twigs and blades of grass that a tiny ant crawls through in his own little universe, which to us may be only a corner of the garden. God has arranged all these different time and distance scales to be appropriate to each species he has created, us included. He is Lord of time and distance.

This is why I find it impossible to accept a literal interpretation of 'the days of creation'. We have to keep reminding ourselves the Bible is not a scientific text book. I believe that God is not to be rushed, and that every stage in the intricate process of creation and subsequent evolution of the universe, from the 'Big Bang' until now, has taken eons of time. We probably all unwittingly limit God, as our brains hurt if we try to comprehend the scale on which he operates, both regarding distance and time. Why should we imagine that God can only work within the confines of our own limited thoughts and understanding?

Although we may find the concept of an infinite God and a limitless universe rather frightening, Christians can take comfort in God's promises. This Almighty God has told us that if we confess our sins he will blot them out completely, and his forgiveness will be absolute.

When God used the phrase

"as far as the east is from the west"
Psalm 103:12

he was referring to the removal of our sins an infinite distance away from us.

And furthermore He says in Jeremiah:

"I will remember your sins no more"
Jeremiah 31:34

So this amazing infinite God even has the power to 'self-delete' some of his memory (even from the recycle bin) and to not only forgive but actually **forget** our sins, once we have repented of them.

When humans forgive each other, even in a thoroughly sincere and well-meaning way, the normal ethic is "to forgive but not to forget" as a sort of safeguard against being hurt again by a recurrence of the problem. By contrast God's forgiveness is so complete he even **forgets** the sin ever occurred! This is an amazing concept, hard for us to take in, but evidently our God is too big to fit into any box we make for him.

*****

1   logarithmic scale – a scale where successive equally spaced divisions are obtained by ***multiplying*** by a fixed number, rather than by ***adding,*** as on a normal ruler.
2   extragalactic – referring to objects in the universe ***outside*** our own galaxy, the Milky Way.

Theme 4   Our Protective Blanket

Sunrise glow in the atmosophere, Kangaroo Bay, Australia

Frank Flynn

> *In his hand is the life of every creature*
> *and the breath of all mankind.*
>
> Job 12:10

Not all planets have atmospheres. To be able to retain an atmosphere[1] a planet must have sufficient mass, otherwise any atmosphere it once had will escape, as the planet[2] does not have enough gravity to hold onto it.

Mercury for example has too small a mass. Its surface very much resembles the Moon. They are both covered with impact craters as there is no atmosphere to shield them from meteoritic[3] bombardment.

Mars is less massive than the Earth but does have a tenuous atmosphere, mainly of carbon dioxide. Venus, which has a similar mass to the Earth, and all the larger planets have dense atmospheres, mainly dominated by carbon dioxide and methane.

Although there is now good evidence for traces of water vapour in various places in the solar system the unique mix of oxygen, nitrogen and water vapour we call **air** is found only on the Earth. Mars once did have a denser atmosphere and significant water, most of which evaporated away millions of years ago, but traces of water have been found to exist, under the surface crust.

The Earth's atmosphere has several key roles in protecting life. The first and most obvious is that we have air to breathe. We also have a water cycle which gives us rain, clouds and a variety of weather conditions. The atmosphere also moderates the harsher effects of temperature and climate, softening the stark contrast of high and low temperature such as is found on the surface of the Moon.

We have blue sky, that is **daylight**, which is sunlight scattered in the atmosphere, enabling us to see clearly, even under a heavy cloud layer when the Sun is not directly present. On the Moon for example there is no daylight. You would see the Sun in a perpetual twilight sky along with other stars.

Don't Put God In A Box

The atmosphere also shields us from harmful radiation, letting through only visible light and radio waves, and protects us from the relentless bombardment of meteors. These become very hot due to the friction of the atmosphere as they pass through, causing them to glow as meteor trails, and most get burnt up before they land. Very large meteors do manage to get through, and cause meteorite[4] craters but these are exceedingly few in number compared with those on the Moon's surface. We will think about this potential hazard to our lives in more detail in theme 8, 'Send a Fireball'.

The list of beneficial properties of the atmosphere goes on. Another one is the propagation of sound waves. Sound does not travel in a vacuum. The Moon is a silent world!

*****

The Lord God made garments of skin for Adam and his wife, and clothed them.
Genesis 3:21

And so certainly the Earth's atmosphere acts like a 'protective blanket', without which life as we know it simply could not function.

We have a loving God who has made infinite provision for the nurture of all life, and given humans the role of looking after the Earth's environment. We are in a position of stewardship, and it is daunting to see so many ways in which we are blindly ruining the Earth's environment. We denude the rainforests and cause climatic changes. We pollute the atmosphere with greenhouse gasses, and cause global warming, with serious consequences to the ecology of the planet.

Hopefully we have not yet reached a 'point of no return' and still have time to pull back without any permanent long-term damage to our planet.

And God said "This is the sign of the covenant I am making between me and you and every living creature with you, a covenant for all generations to come: I have set my rainbow in the clouds, and it will be the sign of the covenant between me and the Earth. Whenever I bring clouds over the Earth and the rainbow appears in the clouds, I will remember my covenant between me and you and all living creatures of every kind. Never again will the waters become a flood to destroy all life'.

Genesis 9:12-15

A rainbow has to be one of the most beautiful natural phenomena, and it has a straightforward physical explanation. It is caused by refraction of light into its component colours by water droplets in the atmosphere at a time when the Sun is shining and it is also raining, so there is water vapour in the atmosphere.

But as with all natural phenomena this too is part of God's creation. It serves as a reminder of God's faithfulness to humankind, and an encouragement when things are not going well, in our private lives, or in the world at large.

*****

1. atmosphere – the gaseous layer surrounding a planet.
2. planet – a major astronomical object in orbit around the Sun. It can be solid (like the Earth or Mars), or mainly gaseous (like Jupiter).
3. meteor – a small solid object attracted towards the Earth by gravity, and caught up in the Earth's atmosphere, seen as a streak across the sky - a 'shooting star'.
4. meteorite – a meteor which penetrates the Earth's atmosphere and impacts the Earth's surface.

Theme 5   The Question of Life

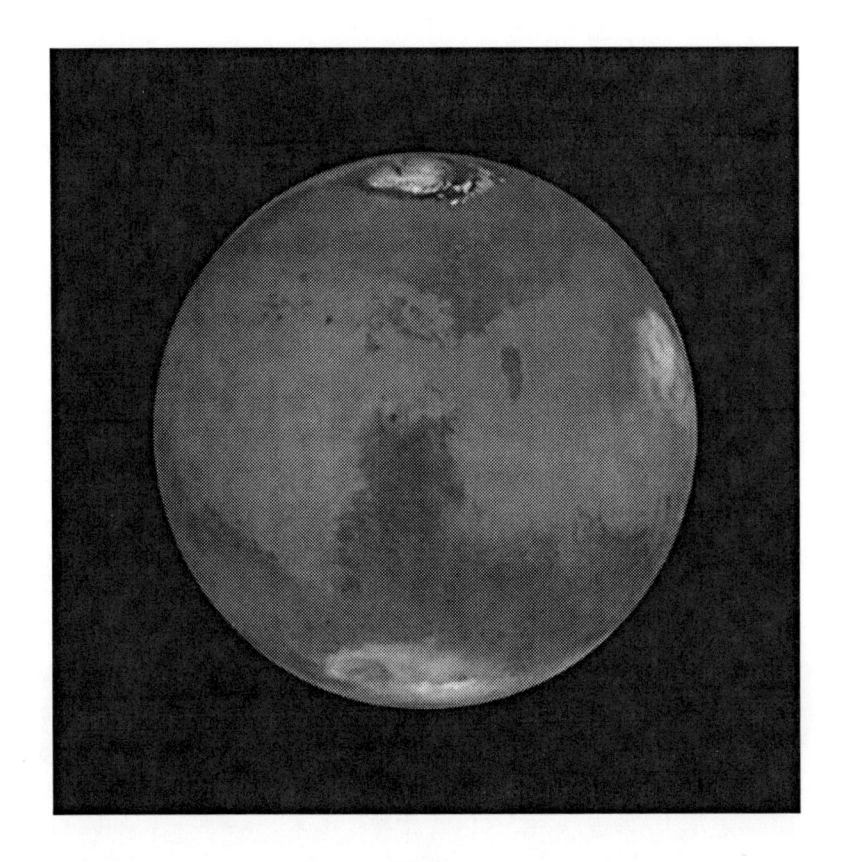

Mars

And God said:
"Let the water teem with living creatures."
"Let birds fly above the Earth, across the expanse of the sky"
"Let the land produce living creatures, according to their kind"
God made the wild animals according to their kinds, the livestock according to their kinds, and all the creatures that move along the ground, according to their kinds.
Then God said:
"Let us make Man in our image, in our likeness".
God saw all that he had made. And it was very good.

Genesis 1

It is only quite recently that astronomy and biology have come together in an interesting fusion called **astrobology**. Astronomers have renewed their interest in the possibility of life 'out there', while at the same time biologists have come to realise that some basic life forms can exist in much harsher environments than previously supposed. These can be exceptionally high or low temperatures, exposure to radiation harmful to humans, very acidic or alkaline conditions, and extreme pressure. They have coined the term 'extremophiles' for these microbes[1]. There is good evidence even on the Earth that life manages to exist in some very unexpected places. 4 km under the ice in Antarctica there is a lake – Lake Vostok, where aquatic life has been found, and also in deep ocean trenches where the pressure is enormous, some primitive life forms manage to exist.

The quest for life in the universe has always been lurking in the background, and there are times when it comes to the fore. There was a particularly striking case at the end of the nineteenth century when the famous Italian astronomer Schiaparelli made line drawings of the surface features of Mars with his telescope in Milan. This was before photography began to be widely used in astronomy, and observers made hand drawings, which were inevitably rather subjective. He used the term '*canali*' for the many linear channels which he observed. The anglo-saxon world misunderstood this Italian word meaning **channels** and interpreted these to be **canals** with the implication they were man-made! Now with detailed close-up pictures of the surface of Mars, we understand these *canali* are great fault lines and rift valleys, giving a linear appearance at a distance but they are purely natural phenomena.

For most of the twentieth century the idea of life elsewhere in the universe fell out of fashion and one was regarded as a bit of a crank for even entertaining the idea. Now however, with a combination of planetary space exploration and the greater realisation of biologists of the existence of extremophiles, the search is on again.

In theme 6 'Are we alone?' we will be looking in particular at this actual search process, but first we should clarify in our minds just what we are looking for. What really *is* life after all? This is a difficult area, requiring careful definition and many people disagree about the criteria. A common definition is **the ability to reproduce,** and so to perpetuate the species. Thus a computer is not 'alive'. But what about a mule? A mule is a cross between a horse and a donkey, and cannot reproduce, so by this definition it cannot be alive either!

*****

There are two sensitive issues about life which often cause difficulty both to believers in God, and non-believers, which should be honestly addressed.

The first is whether or not humankind evolved naturally from earlier species, or whether 'Adam and Eve' were a direct and separate creation by God. This issue of evolution, dating back to Darwin continues to be a difficult and sensitive area. I am not a biologist and would not presume to discuss this from a biological angle. All I wish to say here is that as both a Christian and an astronomer, I do not see the need to adopt a **literal** interpretation of Genesis 1. I am happy with the picturesque and ancient description of the **days** of creation, each implying incredible eons of time. It seems it took God **billions** of years just to process our atmosphere, with the action of bacteria, to produce just the right amount of oxygen to support more advanced forms of life. Why should we hurry God, just because it is difficult for us to comprehend the time-scale in which he is pleased to operate? I don't really see any problem in humankind evolving from earlier species. First there were bacteria, now we have man. Logically somewhere along this chain there had to be a First Man – Adam, created by God 'in His own image'.

The other issue, known as 'biological determinism', is the assumption that if the conditions are right life will **automatically** come about. Thus one might expect, if a planet is discovered elsewhere in the universe, closely resembling the Earth in all its physical properties, life would somehow evolve **as a matter of course**. But many would question this assertion. It is after all a pure assumption. A professor of philosophy, recently talking to a meeting of the Royal Astronomical Society cautioned his audience over the tendency to make this assumption, with the interesting analogy that:

> *'a stick of dynamite (the energy source) under a pile of bricks (the material conditions) does not automatically make a house!*

It is absolutely true to say, at the time of writing, no life form of any type has been found outside this planet. But that does not mean it does not exist. We will go on to examine in theme 6, how the universe is being carefully searched for any evidence of life, past or present.

*****

1   microbe – a minute living being, a micro-organism.

Theme 6   Are We Alone?

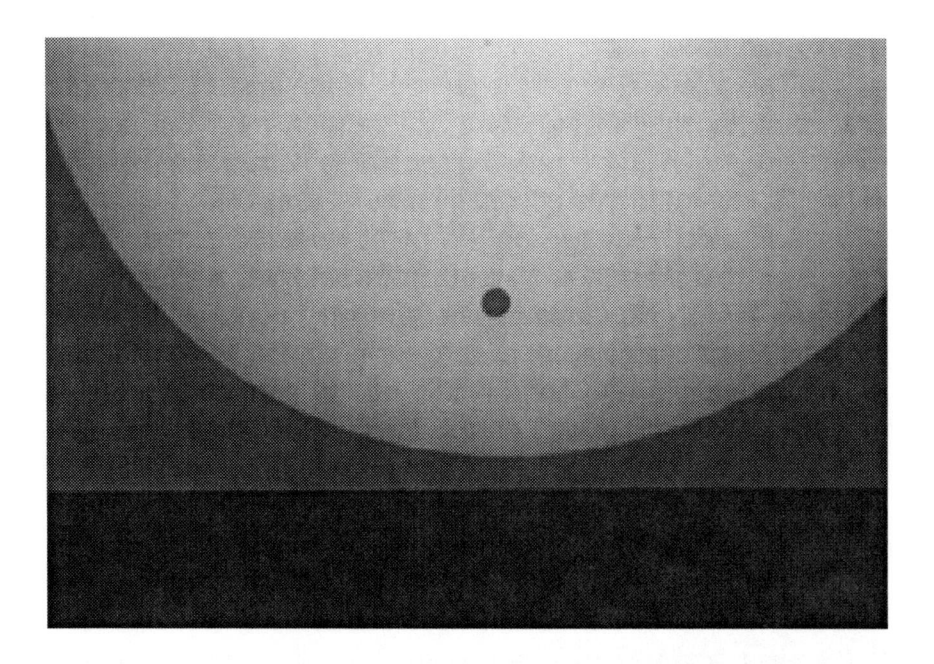

*transit of Venus across the solar disc, June 8th 2004
taken with author's 8" Schmidt-Cassegrain telescope*

Frank Flynn

> *He sits enthroned above the circle of the Earth, and its people are like grasshoppers. He stretches out the heavens like a canopy, and spreads them out like a tent to live in.*
> *Isaiah 40:22*

We will divide our search for life into two parts – first within the solar system, and then in the nearer part of our home galaxy, the Milky Way.

The solar system is the obvious place to start. No longer are we seriously thinking about 'little green men' but, as we considered in the last theme, the possibility of microbes existing in hostile environments now seems more feasible.

Mars continues to be the most favoured place. It has now been extensively surveyed from satellite, and is being explored by robotic landers. There is very good evidence of the past existence of water in some abundance, which caused drainage patterns and flood plains on the surface, not unlike those we are familiar with on the Earth. Most of any residual water is now sub-surface, but some contributes in a frozen state to the 'polar caps' which are seen to increase and diminish with the seasons. It is also thought Mars once had a denser atmosphere, millions of years ago, and was warmer then.

The loss of the British lander **Beagle 2** at Christmas 2003, was a disappointment, although its mother craft the **European Space Express** successfully orbited the planet and has done very useful survey work. American robotic landers have since examined the surface features. Of particular interest is the possibility of discovering fossil[1] samples embedded in the rocks as well as the nature of the planet's surface and its temperature, atmospheric conditions and radiation levels.

There has been great interest in Mars rocks found on the Earth, mainly in the Antarctic. You may wonder how it is possible for Mars rocks to get onto the Earth in the first place! The surprising explanation seems to lie in the role of comets[2]. Comets travel in towards the Sun from the remoter parts the Solar System and are prone to collide with other

Don't Put God In A Box

objects, especially planets. More will be said about the great collision of Comet Shoemaker Levy 9 with Jupiter in 1993, in theme 8 'Send a Fireball'. When such an event happens it is possible material from the impacted surface will be ejected out into space at a speed greater than the escape velocity[3], so that it can actually travel from one planet to another. There was great excitement when it was claimed a fossil was found embedded in one of these Antarctic Mars rocks. However after very careful scrutiny this theory was eventually put on the back burner as inconclusive.

And so although great efforts continue to be devoted to a close search of Mars, no actual evidence of life, past or present, has yet come to light.

There are however two other places of special interest in the solar system, although further a field. One is Jupiter's satellite[4] Europa. This is one of the four 'great satellites', discovered by Galileo in 1610 with the first astronomical telescope. Europa is a similar size to the Earth's Moon and has a complicated fractured surface. But the most significant fact is spectroscopic evidence of a sub surface lake – even a whole sea – which could shelter life forms in a warmer environment, rather like Lake Vostok, 4km under the Antarctic.

The other place of importance in this quest is Titan, the principal satellite of Saturn, and the second largest in the Solar System. It is notable as the only satellite which possesses a dense atmosphere. The Cassini mission to Saturn sent down a probe, Huygens, to land on Titan in New Year 2005. This was an extremely successful operation, and Huygens uncovered fascinating details of a rocky surface and an atmospheric cycle with rain not unlike the Earth's – but this rain is liquid methane not water! Despite striking similarities to the Earth's atmosphere from a dynamic point of view, and surface features such as drainage basins and shore lines uncannily like the Earth's, this is not a 'human friendly' environment. But this does not rule out the possibility of other primitive life forms, and information from Huygens is being studied with much interest at the time of writing.

Frank Flynn

We now look wider – to the neighbouring parts of our own galaxy, the Milky Way. It has always been supposed other stars may have planetary systems, but only recently has this been confirmed by actual observation, and now in 2005, about 150 planets have been discovered. These are known as 'extra-solar planets' or sometimes simply 'exoplanets'.[5]

Stars are so far away it is still impossible to directly see planets orbiting around the mother star, although techniques for achieving higher and higher angular resolution[6] are rapidly advancing. Planets have to be detected by more indirect means. There are several techniques. The most common is by observing a 'gravitational wobble' of the star as it moves in its path through the galaxy. Let's consider our own solar system. When the major planets are all on one side of the Sun, the Sun's centre of gravity[7] is pulled a little out of the centre of the Sun, causing the Sun to slightly oscillate in its path. One could imagine the inhabitants of a nearby solar system observing this wobble, and deducing that the Sun has some mysterious objects with very small mass giving it a little tug first from one side and then the other. The observations are actually done by spectroscopy, measuring very slight variations in the star's velocity in the line of sight.

Another technique is to look for the transit[8] of a planet across the star's disc. Although the star is too far away to be seen as a disc, there will be a tiny drop in the steady light level of the star. This dip is always incredibly small, maybe about one part in a thousand of the light of the star, but it has been detected in a few cases. Anyone who observed the transit of Venus in 2004 will have noticed what a tiny black disc Venus presented against the huge bright background of the Sun, reducing the Sun's brightness by a miniscule amount. Nevertheless such a tiny dip can be observed with sensitive instruments on a very large telescope. It is necessary to observe over a prolonged time to check that this little dip occurs at regular intervals, so we can be sure we really are looking at an orbiting object.

A third technique is known as 'gravitational lensing'.[9] Light is sensitive to gravity, and a massive object will slightly bend a light beam. This phenomenon, important in cosmology in detecting dark matter, can also be useful in looking for planets. If the star in question happens to be directly in front of a bright distant galaxy, the star acts like a lens in bending the light beam from behind slightly towards it. The planet's reflected light from the mother star is hugely concentrated at certain positions in its orbit, and can be detected as a bright flash. This is a new and very sophisticated technique which is still being developed.

So far most of the planets which have been discovered are termed 'Hot Jupiters'. This is because most have masses similar to, or greater than, Jupiter's mass, and yet they seem to lie in inner orbits, more like the position of Earth or Mars in our solar system, so they would be much hotter than Jupiter. Most orbits also seem to be quite elliptical, and more greatly inclined to the plane of the system than in our own solar system where the planets have nearly circular orbits and are nearly coplanar. So although it is now clear 'solar systems' are common in the galaxy, a system closely resembling ours has yet to be found.

Maybe it is not surprising the planets found so far are all big ones. This is a phenomenon called **observational selection.** Imagine again the inhabitants of a planet in a neighbouring system looking towards the Sun and hoping to discover planets. What would they be likely to discover first? Certainly Jupiter and Saturn, and then probably Uranus and Neptune. It might be a long time before our celestial neighbours actually discovered the tiny inner planets Mercury, Venus, Earth and Mars! So the abundance of 'Hot Jupiters' does not exclude the future discovery of other smaller, fainter planets.

In every solar system a region known as the **habitable zone** can be defined. This is basically the zone in which water could exist in a liquid state, at least for a reasonable length of time. In our own Solar System this pretty accurately determines the zone between Venus, and Mars, with the Earth comfortably placed in the middle. There is now much interest in pinning down the location of the habitable zone in other systems. Its actual position in each system depends critically on the temperature of the star. Planet seekers are particularly interested in finding what is sometimes termed an **Earth Clone** planet in the habitable zone. That is, a planet with most of the essential features of the Earth such as a similar mass, temperature and atmospheric conditions.

There is a movement called S.E.T.I. – the **search for extra-terrestrial intelligence.** This is devoted to devising methods for both receiving, and sending messages out into space in an attempt to communicate with any other intelligent beings out in the universe. One could argue mathematically that the probability of there being **Earth Clone** planets somewhere in space is very high indeed, but we should be cautioned by the wise comments quoted in the previous theme that just because the physical conditions are right we should not **automatically** assume life will follow!

*****

In his hand is the life of every creature and the breath of all mankind.
Job 12:10

And so astronomers continue their search for life in the universe. It is a subject of extreme interest and certainly good progress is being made in examining likely **habitats,** whether in our own solar system or further afield in the galaxy. However life has not yet been found outside this planet, in any shape or form.

Scientists are usually good at keeping an open mind about situations that have not yet been proved or disproved. I believe as a Christian I too should keep an open mind on this subject. Once again it is all about the un-wisdom of limiting God. Three points of view come to mind – two extremes, and one in the middle.

The first is the view that life – at least **human** life – has to be unique to this planet. Some Christians would quickly assert that the Bible gives no clue about life anywhere else in the universe – true, but does this omission necessarily imply that it does **not** exist? I would interpret the Bible's silence on this subject to be that a knowledge of 'yes or no' is simply not relevant to our salvation. We have to remember yet again that the Bible is not a text-book of science.

The second view, at the other extreme, is the concept of **biological determinism**, mentioned in theme 5. This is that life would **automatically** arise in any situation where the conditions are right, as a perfectly natural physical process. It is rather difficult for a Christian to subscribe to such a view, which effectively rules out the unique hand of God the Creator, and simply turns life into another law of nature.

But there is a third, intermediate view, which harmonises both the uniqueness of God the Creator, and the rather exciting possibility of plurality. It is simply to respect the omnipotence of the God of the Universe. Why should God not choose to have families on other planets too if he wants to? If God has in fact created life elsewhere in the universe, this does not in any way undermine our own relationship to him. God is absolute sovereign, and he will do what he will.

Some very thorny questions do however arise, such as The Fall. Are there really any 'perfect' worlds out there? And what would happen about the redemption of another sinful planet? Of course we cannot possibly presume to say, but it does highlight the serious responsibility of would-be space explorers and the question of contamination of the universe. Physical contamination at least is taken seriously, and there is a very well developed consciousness in the space community about 'clean' space missions and the need to avoid any form of environmental pollution. 'Moral' contamination is however a different matter, and raises huge ethical questions about the colonisation of space, if indeed this day should ever come, and the relationship of sinful earthlings to any other life forms encountered elsewhere in the universe.

*****

1   fossil – traces of the remains of an ancient animal or plant, hardened and usually embedded in rock.

2. comet – a body made up of frozen gasses and solid particles, orbiting the Sun, generally in a very extended elliptical orbit,. The gasses vaporise when near the Sun, forming a diffuse gaseous head surrounding the comet's solid nucleus, and a tail.
3. velocity – the speed of an object in a given direction, e.g. in the line of sight.
4. satellite – a secondary object rotating around a primary, e.g. the Earth's Moon.
5. exoplanet (=extra-solar planet) – a planet orbiting a star other than the Sun.
6. angular resolution – the degree to which two distinct objects can be seen separately in angle.
7. centre of gravity (= barycentre) – the common point about which a system of bodies rotates.
8. transit – when a secondary body moves across the face of the primary, e.g. Venus transiting the Sun.
9. gravitational lensing – the bending of light rays due to the gravitational attraction of a massive object, acting rather like an optical lens.

Theme 7   Man's Conceit and Impotence

The constellation of Orion

Frank Flynn

> *"To whom will you compare me ? Or who is my equal?" Says the Holy One. "Lift your eyes and look to the heavens. Who created all these?" He who brings out the starry host, one by one, and calls them each by name. Because of his great power and mighty strength, not one of them is missing.*
>
> *Isaiah 40:25-26*

> *Hear and pay attention. Do not be arrogant, for the Lord has spoken.*
>
> *Jeremiah 13:15*

Running through the long history of man's increasing understanding of the universe, there has always been a tendency to give ourselves 'pride of place' – to place ourselves at the centre of things. This tendency is so engrained in us it is almost subliminal. There is a psychological thread running right through the history of astronomy which I call '**centerism**'.[1] Here are three examples.

Firstly, and probably the most significant event in the whole history of science, was when Galileo finally established that the Sun did **not** revolve round the Earth, but the Earth round the Sun. Now, 400 years later, we take this simple fact for granted, but not so in seventeenth century Catholic Italy. This is not the place to go into detail about this fascinating and defining period. Copernicus had earlier produced his **heliocentric**[2] theory, that the planets revolve around the Sun in circular orbits. But this was simply a theory, a mathematical model. It seemed to greatly simplify the complex system of **epicycles**[3] first put forward by Ptolemy in about AD 100, to account for the planets moving in strange 'zig-zag' paths as they apparently rotated about the Earth, along with the Sun. The Copernican system was not too much of a threat as long as it was simply taken as a mathematical model, and not as **reality**!

In 1610, with a primitive telescope, the first ever used in astronomy (which can still be seen in the Science Museum

Don't Put God In A Box

in Florence), Galileo discovered the four principal moons of Jupiter. He saw they were actually orbiting Jupiter itself, and not the Sun. This was the first time any such mechanism had been observed, running counter to the established wisdom that the Earth was the pivot of the Universe. This discovery was like touching a live wire. It threw the establishment into turmoil. When Galileo also observed that the planet Venus exhibited phases like the moon, due to light reflected from the Sun at different angles, and also periodic changes in the *size* of the planet's disc, it became clear Venus too was orbiting the Sun, not the Earth.

It took many years of intrigue and persecution, both of Galileo himself and other free thinkers before the religious establishment of the day eventually conceded that after all God had actually ordained that the Earth went round the Sun. And so humankind finally lost its 'pride of place' in the universe.

We should remember Galileo was himself a devout Catholic, and not in any way trying to undermine the faith. His only 'sin' was to use the mind that God had given him, to think, and be honest. So now for over 400 years we have had to eat humble pie, and to get used to our more modest position in the astronomical hierarchy as dwellers on a rather small planet, along with other mightier siblings, orbiting our parent star, the Sun.

Now we come to the second case of 'centerism', our place in the galaxy. But rather we should say ***the Sun's*** place in the galaxy, as on this scale, we the Earth and all the other planets of the solar system are effectively subsumed into the Sun itself, as a mere dot in space.

There was little understanding of the nature of our own galaxy, the Milky Way, until the time of Sir William Herschel, in the late eighteenth century. Herschel investigated **stellar parallax**[4] which enables us to calculate the distances of stars. He began to build up a picture of the shape and size of the Milky Way – our own home galaxy. We now know that the Milky Way is a spiral galaxy. Its plan view presents a spiral shape, rather like a gigantic 'Catherine Wheel' firework, and edge-ways on it is a flattened disc-like structure, with a central bulge – like two saucers placed one on top of the other.

As we look around us in the night sky, well away from any light pollution, we can see this faint milky band straddling the whole sky. It is tempting to think we must be in the middle of it! Once again the psychology of 'centerism' comes into play. Just because this milky band appears to straddle us, does this automatically place us at the centre? No, it does not. We can be at a random position somewhere in the plane of the galaxy, and as we look around us in the plane we see a greater concentration of stars than we would at right angles to the plane. We are somewhere in the plane of the galaxy, but not necessarily at the centre.

Astronomers became a little more cautious, and happier to accept that 'we' – that is the Sun – are not necessarily at the centre of the system. Now we find we are about two thirds of the way from the centre to the edge, in a rather random position in the galaxy. Man's pride had taken another knock.

Now we come to the third case of 'centerism'. Looking outwards from the Milky Way, our home galaxy, we see that the light from distant galaxies is 'red-shifted'[5] – that is that the wavelength of the light is increased, and the objects are receding from us. This is a fundamental cosmological fact. Galaxies seem to be moving away from us, and evidently the universe is expanding.

Does this put us at the centre of the universe? No, it certainly does not. We considered in theme 1 that If you imagine dots painted on a balloon which is expanding, every dot is moving away from every other dot. There is a **mutual** expansion of the whole system. Just because we see everything moving away from **us** does not make us the centre of the system. We are merely a member of a small group of galaxies, sharing in the general cosmological expansion We are evidently in a **random** position in the universe too.

*****

These three historical experiences of losing our 'pride of place' – first in the solar system, then in the galaxy, and then in the universe overall, have taught us a lesson. God has **not** placed us in any sort of privileged position in the universe.

Why then should mankind be so 'puffed up' with his own importance?

From a Christian point of view this has to be due to sin. We get a clue from the serpent whispering to Eve in the Garden of Eden:

> *God said "You must not eat fruit from the tree that is in the middle of the garden, and you must not touch it, or you will surely die".*
>
> *"You will not surely die" the serpent said to the woman, "for God knows that when you eat of it your eyes will be opened, and you will be like God, knowing good and evil"*
>
> *When the woman saw that the fruit of the tree was good for food and pleasing to the eye, and also desirable for gaining wisdom, she took some and ate it. She also gave some to her husband who was with her, and he ate it.*
> *Genesis 3:3-6*

And so pride entered into man's heart and he thinks he can be 'like God'. Certainly, as he is evidently so important, he assumes God would naturally have given him a privileged position in the universe. This has been a hard lesson to unlearn.

Coupled with this inflated sense of his own ego in the order of things, there came an arrogance, a feeling that man could do everything, and achieve everything, given only a little time. Soon after these events as described in Genesis comes the story of the Tower of Babel:

> *"Come let us build ourselves a city, with a tower that reaches to the heavens, so that we may make a name for ourselves.........."*
> *Genesis 11:4*

But God had other ideas and exposed our complete impotence at a cosmological level in his dialogue with Job:

> *"Where were you when I laid the Earth's foundation? Tell me, if you understand. Who marked off its dimensions? Surely you know! Who stretched a measuring line across it? On what were its footings set, or who laid its cornerstone?"*
> *Job 38:4-6*

> *"Can you bind the beautiful Pleiades? Can you loose the cords of Orion? Can you bring forth the constellations in their seasons, or lead out the Bear with its cubs?"*
> *Job 38:31-32*

We may feel somewhat chastened, but I am sure this does not mean we should stifle our intellect. God has given us minds to think with and expects us to use them honestly and to the full. I believe we should learn from these ancient biblical lessons and temper our intellectual pursuits in studying the wonders of the universe with an appropriate humility when we consider the greatness of our creator and the smallness of our position in the order of things.

However to end this somewhat depressing theme on a more cheerful note, we have good evidence from the Bible about how God has a habit of using 'small things', and 'small people' to do incredible things. We will consider this further in theme 9, 'Small is beautiful'. However before this, in the next theme 'Send a fireball' we first have to look at another worrying problem – how God, if He wished, could so easily put an end to our existence – and how several times in even quite recent astronomical history this has very nearly happened.

*****

1. centerism – the psychological tendency to automatically place ourselves at the centre of our particular 'frame of reference'.
2. heliocentric – the system in which the planets orbit around the Sun, as distinct from geocentric – all objects orbiting the Earth.
3. epicycles – a system of circles within circles, devised by Ptolemy, to explain the strange zig-zag, or **retrograde,** motion of the planets. (An **epicycloid** is the mathematical curve traced out by a mud mark on a bicycle wheel rolling down the bottom a hill and up the other side).
4. stellar parallax – a method of finding distance by observing the position of a star on two occasions six months apart, so that the Earth has moved to the diametrically opposite side of the Sun in its orbit, and we are looking at the star from two slightly non-parallel directions. The **parallax angle** measured is exceedingly small – less than a second-of-arc even for nearby stars. (60 seconds of arc = 1 minute of arc, 60 minutes of arc = 1 degree).
5. red shift – the spectrum lines of a star or galaxy are slightly **increased** in wavelength, i.e. shifted towards the red end of the spectrum, due to the object moving away from us in the line of sight. (blue shift is the reverse, a slight decrease in wavelength due to the object moving towards us).

Theme 8 Send a Fireball

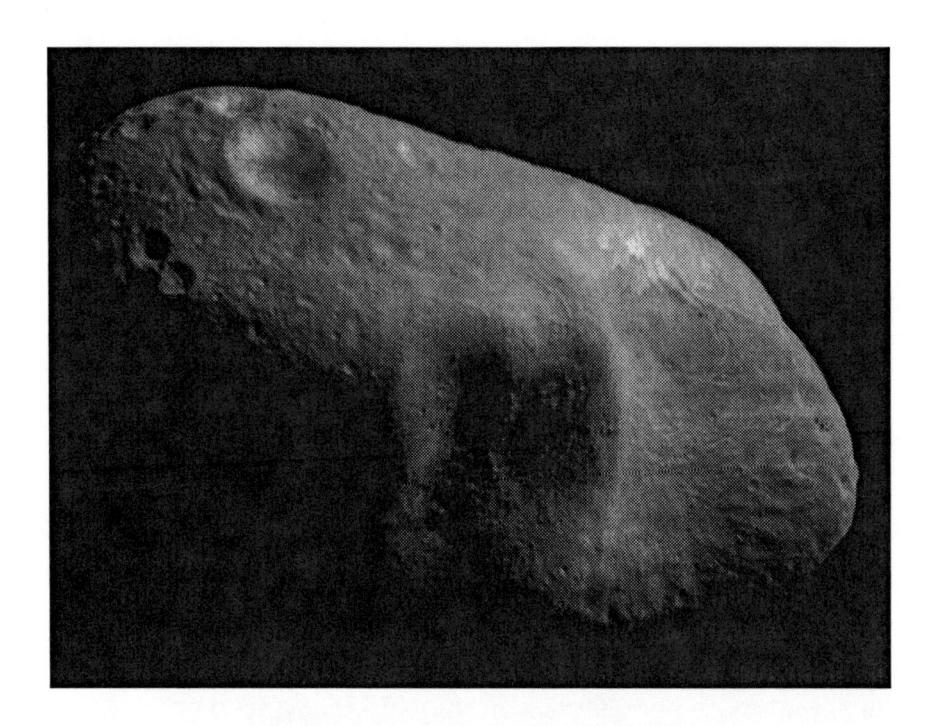

The asteroid Eros from the spacecraft NEAR
The long axis of this asteroid is about 35 km

The third angel sounded his trumpet, and a great star, blazing like a torch, fell from the sky on a third of the rivers and on the springs of water.
Revelation 8:10

"When your fear cometh as desolation and your destruction cometh as a whirlwind, when distress and anguish cometh upon you, then shall they call upon me, but I will not answer, they shall seek me early but they shall not find me. For they that hated knowledge and did not choose the fear of the Lord, they would none of my council, they despised all my reproof. Therefore shall they eat of the fruit of their own way, and be filled with their own devices".
Proverbs 1:27-31 (AV)

In July 1994 Comet Shoemaker-Levi 9 hurtled into Jupiter, the greatest head-on collision in recorded history. This comet had only been discovered a short time before. It had run too close to Jupiter on its journey round the Sun and had been 'captured' by Jupiter and so was now rather unusually a satellite comet of Jupiter, moving round Jupiter in an elliptical orbit. It was clear it was moving in dangerously close to Jupiter itself and in its penultimate orbit in 1992 it passed so close to the surface of the planet that Jupiter's massive gravitational force shattered it into fragments.

About 20 large remnants continued in orbit, each a 'mini' comet complete with tail. The largest fragment was about 5km across. They came to be known as the 'string of pearls'. So much kinetic energy[1] was lost from the Jupiter-grazing encounter it became clear this comet would not be able to escape and this would be the final orbit. When the 'string' again came in close to Jupiter it was certain each component would suffer a head-on impact and become annihilated. Comet specialists were able to calculate closely the expected place and time of the impacts of each member of the string.

Don't Put God In A Box

As fate would have it the impacts all occurred round the back of Jupiter as seen from the Earth. However Jupiter has a very rapid rotation period of about 10 hours and the evidence of these massive impacts soon became visible. Jupiter was pockmarked with impact 'scars' stretching in line over a large part of its surface. Because Jupiter is a fluid object these collisions were really 'splash-downs. The gas ejected from each impact shot up into space and then splashed down again like giant fountains covering wide areas up to 1000 km in diameter.

To get some idea of the scale of this impact, imagine a huge rock 5km in diameter hurtling towards the Earth with a speed of 80 km/sec! It is not surprising that this event marked a defining moment in our attitude towards possible Earth collisions. Until then it had been known that occasionally a large meteor, or possibly even a small comet or asteroid did in fact encounter the Earth, but no one in the astronomical community or elsewhere really took the threat very seriously. However since Shoemaker-Levi this subject has been treated with considerably more respect, and now many observatories have staff whose job it is to keep a watch for intruders. Likely candidates are mainly meteors but could also be asteroids which have strayed away from their habitual orbital belt between Mars and Jupiter. They could even be rogue comets coming in close to the Sun, and passing uncomfortably close to the Earth.

Meteors have already been mentioned briefly in theme 4. They are mainly small rocks and boulders and the Earth's atmosphere burns them up. Larger ones do penetrate the atmosphere and land, and there are some famous meteorite craters such as the Barringer Crater in Arizona, a mile across. But there is also a rather comforting graph which connects the size of the intruders to their frequency of arrival – the bigger the less frequent. However, although those large enough to cause serious damage are relatively infrequent, the historical statistics are not altogether reassuring. The Earth has suffered some very serious knocks in the past.

In 1908 a small asteroid or comet hit the Earth in the remote Tunguska River area of northern Siberia. This is known as the 'Tunguska Event'. Due to the extreme remoteness of the spot little was realised about the scale of this impact until years later. Evidently the object exploded like a fireball [2] before landing due to the extreme pressure of the atmosphere breaking its fall and causing enormous heating. The blast of the explosion devastated 3000 square km. of forest. This is an example of an 'intermediate level event', which can potentially cause very serious localised damage and has a frequency of probably every few hundred years. Luckily in the case of Tunguska very few people were killed, or even injured, due to the remote location. Had the epicentre been London, little of the city and its inhabitants would have survived.

At the 'extremely serious event' end of the graph the frequency is now measured in millions of years. There is very good geological evidence of a massive Earth impact, the 'Yucatan Event', when some 65 million years ago a major object, probably a comet, collided with the Earth in the Yucatan area of the Gulf of Mexico. The resulting devastation led to thousands of years of 'nuclear winter' on the Earth as the debris of the impact polluted the atmosphere and cut out the Sun. This almost certainly led to the demise of the dinosaurs and other major species of the time, and it took the Earth thousands of years to recover. So although such an event is extremely infrequent from the point of view of our life span, or even the course of human history, we have to accept that it **can** and **does** happen, on an astronomical time scale, from time to time.

*****

He causes his sun to rise on the evil and the good, and sends rain on the righteous and the unrighteous.
Matthew 5:45

Don't Put God In A Box

Atheists would argue these events are all part of the natural order. Humans can do nothing about them, and there is no God to intervene, even if he wanted to.

The whole question of natural disasters is very difficult to comprehend and one which we should not be glib about. Believers and non believers alike were left shattered by the scale of the tsunami disaster on Boxing Day 2005. The tsunami seems to fall within the category of a totally 'natural' disaster. There is no way the hand of man can be linked to a geological phenomenon which is a direct result of the planet still being in the process of cooling down.

I believe there is a loving God, who does not deliberately inflict disasters on mankind as a sort of punishment. On the other hand he clearly does allow natural disasters to take place. He has actually placed us on a rather dangerous planet which is still geologically active, and it is clear he gives no preferential treatment to believers.

I would not subscribe to the totally naturalistic argument however that God, having set the universe in motion, then abdicates from any further direct influence on events. I believe God does sometimes perform miracles – that is divine interventions into the natural order – for very specific reasons, but most of the time he works through the natural laws of physics he created in the first place.

Humans have made such a dreadful mess of the world, and of their own lives that one might imagine God would have had enough of it. He could have 'sent a fireball', and finished the whole thing off. We are only too aware how easy this would be. The Old Testament paints a vivid picture of God, often depicting him as an almost 'human' personality, constantly striving with man and giving him a second chance. This comes through very powerfully in the New Testament, where we see God's love for man so clearly in the sacrificial death of his own son Jesus, taking all the sin and the wrongdoing in the world onto his own shoulders, so that we could live.

We cannot rule out the possibility of future natural disasters, great or small, but we do know we have a God

Frank Flynn

whose patience with us is evidently as great as the scale of the universe itself.

*****

1 kinetic energy – the energy a body has due to its ***motion.***
2 fireball – a meteor, or other incoming object, which explodes before landing due to the extreme pressure of the Earth's atmosphere acting as a break and causing enormous heat.

Theme 9   Small is Beautiful

The Earth
first seen as a planet from Apollo 11

Frank Flynn

> *'The kingdom of heaven is like a mustard seed, which a man took and planted in his field. Though it is the smallest of all your seeds, yet when it grows, it is the largest of garden plants and becomes a tree, so that the birds of the air come and perch in its branches.'*
>
> *Matthew 13:31-32*

One of the things about astronomy which never ceases to amaze is the mind-blowing scale of the universe, both in distance and time, as we have seen earlier. Astronomers have not got 'priestly powers'. They only manage to cope with the dimensions involved by a crafty system of 'scale modelling', so that each layer of the universe is effectively reduced to a dot before considering the next zone outwards. Thus we reduce even the Earth-Moon system to a dot in space before considering the solar system as a whole, and then the whole solar system is subsumed into the Sun when we consider our own galaxy, the Milky Way. Then again the Milky Way, and other whole galaxies are effectively reduced to mere dots in space when we confront the universe as a whole.

The same is true of the time scale of the universe. Light takes a mere second to reach us from the Moon, but a hundred thousand years to just cross the diameter of the Milky Way galaxy, and up to thousands of millions of years from the most remote objects visible in world-class telescopes!

Strictly in terms of physical dimensions the Earth is a microscopic dot of a planet associated with a very modest star, the Sun, which is one of some hundred thousand million other stars in the Milky Way galaxy. The Milky Way itself is only a modest galaxy amongst the hundreds of millions of other galaxies in space.

We are like that **'grain of mustard seed'** referred to above.

*****

> *'But God chose the foolish things of the world to shame the wise; God chose the weak things*

Don't Put God In A Box

of the world to shame the strong. He chose the lowly things of this world and the despised things – and the things that are not – to nullify the things that are, so that no one may boast before him'.
1 Corinthians 1:27-29

On the face of it our insignificant position within the mighty universe is chilling, and the prospect of the Earth being the only planet sustaining life is a very lonely thought. Perhaps this explains such compulsive interest and effort going into the search for life elsewhere, which we considered in themes 5 and 6. In the final analysis our extreme smallness and isolation in the universe is simply a fact of life to be accepted. We can't do anything about it. However Christians derive comfort from an understanding that the small things of creation, and small people, are precious to God and are often used by Him in big ways.

Often in Old Testament days it was the youngest and least significant brother in a family who was called to high things. David, a little shepherd boy and very much the youngest and least important of the sons of Jesse, became the most famous King of Israel.

The Bible is full of examples of how 'small people' are absolutely pivotal to God's purposes. We remember in the New Testament how God chose that humble but obedient Christian Ananias to greet the feared persecutor of the church, Saul, resulting in Saul becoming the apostle Paul – the greatest missionary of the early church.

Jesus too was indignant when the disciples regarded little children a nuisance and wanted to send them away, and to their surprise he gathered them together and made a fuss of them.

And so we do not need to mourn our lowly position in the universe.

God loves small things, and small people. Small is beautiful.

Theme 10   Man in Space

Buzz Aldrin on the Moon. June 1969

Frank Flynn

> *'Canst thou by searching find out God? canst thou find out the Almighty unto perfection? It is as high as heaven; what canst thou thou do? deeper than hell; what canst thou know? The measure thereof is longer than the Earth and broader than the sea'.*
>
> *Job 11:7-9 (AV)*

In 1957 Sputnik 1[1] went into Earth orbit, and heralded the start of actual space exploration, as distinct from passive observation of the universe from the Earth through a telescope.

Since then there have been so many satellites, space probes and landers it is difficult to keep track. The Russians were first with a man in space, but since then the Americans have acquired a strong lead in the whole field of manned and unmanned space exploration. The European Space Agency now also has a prominent role, and most other developed countries have a stake in space of one sort or another.

The days of the 'space race' are now really over, and space exploration is happily a much more collaborative affair. Launching space craft, and funding the sophisticated instrumentation aboard is massively expensive, and worthwhile missions can only be achieved by serious international cooperation.

We should distinguish between **unmanned** space exploration, which is by far the biggest sector, and the **manned** programmes. These have so far been limited only to Earth-orbit expeditions, such as the manning of orbital space observatories, and the Apollo[2] programme to the Moon. It is important to realise that although unmanned space technology is now so advanced, man himself has so far hardly left this planet.

Probes were first sent to the inner planets Mercury, Venus and Mars, and then in 1977 Voyager 1 and 2 were launched to take account of a favourable alignment of the four giant outer planets, Jupiter, Saturn, Uranus, and Neptune. The Voyagers were phenomenally successful in making close approaches to these planets for the first time ever. Since then there have been many more sophisticated missions, probes, and landers to many destinations in the solar system. There has been special interest in Mars, and more recently also asteroids and comets. The last major achievement has been to actually place a lander, Huygens, on the surface of the principal satellite of Saturn, Titan.

All these missions have led to an information explosion about the solar system. Our factual knowledge of the solar system is increasing exponentially – doubling about every ten years.

These space projects are enormously expensive, but of course nothing in comparison to sending humans into space, and the technology required to bring them home again. So what of the prospects for **manned** space expeditions?

There is an interesting debate in astronomical circles about whether a human can function more efficiently than a robot in space, in exploring the surface terrain of say the Moon or Mars. Arguments seem to be quite finely balanced. But in the final analysis the whole question of human space exploration belongs more to the realm of the human spirit, rather than to science in a 'clinical' sense.

*****

'Where can I go from your spirit? Where can I flee from your presence? If I go up to the heavens, you are there; if I make my bed in the depths, you are there'.
Psalm 139:7-8

"Why climb Mount Everest?"
"Because it is there".

The human spirit is such that we all have an inbuilt need for fulfilment. In adventurous people this takes the form of striving to extend the boundaries. In 1911 Amudsen and

Frank Flynn

Scott vied to be the first to the South Pole. In 1953 Hillary and Tenzing were the first to climb Mount Everest. In 2005 Ellen MacArthur won the ladies' record for sailing round the World single handed. The list of 'firsts' is endless, and now extends into space.

Some Christians take the line that a lot of these adventures – especially the space-bound ones – are achieved at enormous cost, and the money could be better spent on the humbler task of relieving so much misery, poverty and deprivation in the world today. This is a very reasonable argument, but difficult to resolve. We recall the disciples confronting Jesus about the woman 'wasting' so much money anointing him with expensive perfume, when that money could have been given to the poor. Interestingly Jesus strongly refuted this, saying **"The poor you will always have with you".** It is all a big balancing act.

I actually believe God has placed a questing spirit deep inside us. It is not wrong. It is one of the things which makes us human, and that in pursuing our talents to the full, we are glorifying God. We are not all great explorers. Most of us have much humbler objectives, but there is place for the 'big time' as well, and this includes space travel. Next project: **Mars.**

The quest for life on Mars has already been mentioned in theme 6 'Are we alone?' Many satellites have been sent into Mars orbit, and several landers, and there are now plans for a manned expedition to Mars. This will be a long and lonely voyage. A Moon visit amounts to a 10-day round trip, but to travel to Mars and return, even when the orbital conditions are most favourable, will be well over a year's journey. As well as the technical aspects of such a journey there are also enormous physiological and psychological problems to overcome. How would you like to be stuck with one or two other people in a tiny space capsule for well over a year with no possible chance of getting out? It does seem likely however that such an adventure could feasibly take place within the next 20 years or so.

In part this will be to fulfil scientific objectives – to explore the surface of the planet in much the same way as robots are already doing, but with the additional advantage of human

Don't Put God In A Box

intelligence and the possibility of flexibility – of making tactical decisions and changing plans if it is advantageous to do so. But the other aspect has to be 'pure adventure', simply an extension of Earth-bound adventures such as going to the South Pole, or climbing Everest.

There is also the question of colonization – could colonies of earthlings live for long periods on say the Moon, or Mars? Serious research is being done into such questions. The subject of **terraforming** is concerned with the problem of developing and controlling environments (like the Eden Project on a much bigger scale) in which humans could actually live and lead a quasi-normal existence, by adapting local physical conditions such as atmosphere and climate. This is a fascinating study, and simulations are being tried on the Earth. But while a certain degree of 'living off the land' is probably possible, it is difficult to see how this could be done effectively on a large scale, or for long periods of time. There are plans for a permanently manned observatory on the Moon, to act as a relay-station for a manned visit to Mars. This would however still be a 'space-station', not unlike an orbiting space observatory, totally sealed off from the local environment, and where workers would come and go for limited periods of duty. This in itself is not a step to long-term colonization.

While Mars is certainly feasible from the point of view of manned missions, it would be a massive further step to go anywhere beyond this. The purpose of journeys beyond Mars would be difficult to justify. There is really nowhere else to go where it would be safe to land. Journey times would be immense, and they would almost certainly be one way trips. And this is just within the solar system itself. To travel to even the nearest planetary systems beyond would not only be a one way trip, but would take many generations to accomplish, with complete life support systems on board!

So the space-age has dawned. Some limited human space travel has already been accomplished and is likely to be extended within the inner regions of the solar system in the comparatively near future. But most of us will have to be content

Frank Flynn

to live on the planet that God has provided, enjoy it, and try very hard not to wreck it.

*****

1   Sputnik – a series of 3 unmanned spacecraft, launched into Earth orbit by the Soviet Union, first in 1957.
2   Apollo – an American space programme to send spacecraft to the Moon. There were 17 Apollo missions in all. Nos. 1 to 6 were unmanned test flights, There were 6 Manned landings. The first was in July 1969.

Theme 11   Star of Bethlehem

Comet Hale-Bopp 1996, with two tails

Frank Flynn

> *Magi came from the east to Jerusalem and asked, "Where is the one who has been born Kind of the Jews ? We saw his star in the east, and have come to worship him".*
> *Matthew 2:1-2*

> *After they had heard from the king, they went on their way, and the star they had seen in the east went ahead of them until it stopped over the place where the child was. When they saw the star they were overjoyed.*
> *Matthew 2:9-10*

Much has been written about the famous "Star of Bethlehem" over the years. It is a story which never ceases to fascinate. Just what **was** this 'star'? As well as its significance in the bible story, it deserves attention in its own right as an astronomical phenomenon.

There are several serious theories about what it could have been, and papers have been written expounding the virtues of one against another. Assumptions have to be made, and there are many ways of interpreting the evidence, so there is no final consensus. We are not going to attempt a detailed study of the subject here, but just briefly consider two of the more prominent ideas.

The first that of a **planetary conjunction**[1]. As planets orbits are tilted only very slightly to the **ecliptic plane**[2], they all move in nearly the same line in the sky. As they are moving at different speeds, one will sometimes pass behind another as seen from the Earth. This is called a conjunction. However, although we think we can only see one object, it is extremely unlikely in fact that one will exactly obscure the other in our line of sight. This is because the disc sizes presented by the planets in the sky are such a small size the planets are effectively like two separate dots close together in the sky each emitting its own shaft of light. Our eye cannot resolve the two separately and so we see only one shaft, but we get the benefit of the

Don't Put God In A Box

light from the two planets added together. So the result is a much brighter image than we would normally attribute to one single planet. Venus and Jupiter are normally the two brightest planets. If their light were combined we would see an immensely bright single point of light in the sky, for a limited time. This would certainly be a spectacular sight.

The weakness of this as an argument for the 'Star of Bethlehem' however is that the two separate planets would have been seen a little time before and after the conjunction. To seasoned observers, as these Magi must have been, the reason for the event would have been rather obvious, as they would see the planets gradually coming together over a period of a few days. It would still however be very unusual. They would almost certainly not have seen such an event before, and the possibility of some divine happening would capture their minds.

I am not too keen on the planet conjunction idea. It is just a bit too **ordinary**. I prefer the other main alternative – the visitation of a **bright comet**.

Comets have been mentioned briefly in theme 6 'Are we alone?' Throughout history they have mystified mankind. As their orbits are mainly very elliptical they only come close to the Earth and the inner planets when they are also in the vicinity of the Sun. This is when the gasses in the head of the comet are shining most brightly by a process known as **florescence**[3]. This mechanism is very sensitive to the comet's distance from the Sun. There is usually a short, sharp flare up of brightness as the comet approaches the Sun, and then an equally steep drop as the comet moves away from the vicinity of the Sun only a few days later. This explains why comets appear to be such elusive objects and we generally only see them for very short intervals of time, although their orbital **periods**[4] may be thousands of years.

Newly discovered comets are often a sensation, especially if they are bright. Comet Hale-Bopp, in 1997, was particularly spectacular, with its brightly shining **coma**[5] and two distinct tails. It was with us for several weeks and became a good friend in the sky each night before it gradually decayed in brightness as it moved away from the Earth and the Sun, to continue its orbital period of about 4000 years before visiting us once again.

Although comets are all believed to be periodic – that is they have elliptical orbits round the Sun – many have such long periods there is no hope of seeing them twice, and when first observed they rank as 'new discoveries' although they have in fact visited the Earth long ago.

The Magi would have been spellbound to see the apparition of a new bright comet. It would have moved slowly across the sky from the east to the west during the night as the Earth rotated, setting gradually in the west – the direction to which we assume they were travelling. Each night it would change its position in the sky a little before it set, until after only a few days it would fade away for ever.

I would not interpret the 'stopping' phrase in Matthew's account absolutely literally, but I can understand that it could give the impression of standing still in the sky as it gradually set in the west, acting each night like a pointer. A comet's tail points away from the Sun, and the head and tail combined could resemble something like an arrow, pointing downwards into the sunset in the west, where Bethlehem lay before the travellers.

*****

I feel very much that God, the King of the Universe, was deliberately using an astronomical sign to lead the Magi to where Jesus was. We know this event was probably a year or two after the birth, and so not a visitation to the manger itself, but nevertheless still to Bethlehem.

In looking for an astronomical explanation we are not necessarily ruling out the miraculous, but there is no reason why God should not choose to use a perfectly natural although unusual event as a sign.

Don't Put God In A Box

If this is the case it is fascinating to see once again how God is the God of time, and never to be rushed. If the 'star' were in fact a comet, God must have requisitioned that very comet for its special purpose, thousands of years earlier, so that it would have its ***perihelion passage***[6] at exactly the right time to guide the Magi at night, and go down in history to mark the most important nativity of all time. It gives us a powerful insight into how God is in control of historical events as well as every detail of the dynamics of the universe.

As far as we know comets are all periodic, that is they move in elliptical orbits and do eventually return, although in the case of comets with very long periods, their return is a matter of many thousands of years. When Comet Hale-Bopp last visited the Earth it must have been about the time of Stonehenge, or Moses.

When a new comet is discovered the first thing astronomers do is to try to pin down the details of its orbit and work out its period. What if one day someone discovers a 'new' comet with a period of about 2000 years? Would we be really looking at the original 'Star of Bethlehem' returned?

*****

1. conjunction – when one object moves behind another in the line of sight.
2. ecliptic plane – the plane of the Earth's orbit around the Sun, generally taken to be the fundamental plane of the solar system.
3. florescence – the process by which a comet will flare up in brightness when near the Sun, by the gasses in the comet's head absorbing sunlight and then re-emitting it, like a florescent watch dial shining in the night.
4. period – the length of time for a planet, or a comet to complete one orbit about the Sun.
5. coma – the gaseous head of a comet.
6. perihelion passage – the point in time when the comet is closest to the Sun.

Theme 12   Total Eclipse

*Total eclipse of the Sun
11th June 1991. La Paz*

Frank Flynn

> *If I say "Surely the darkness will hide me and the light become night around me," even the darkness will not be dark to you; the night will shine like the day, for darkness is as light to you.*
>
> *Psalm 139:11-12*

Of all astronomical phenomena, a **total eclipse of the Sun**[1] has to be the most striking and evocative. I have been lucky enough to see a total eclipse twice, and both times it has left a deep imprint on my mind. As well as the visual impact there are profound psychological aspects. First let's think about what is actually happening.

The Sun's distance from the Earth is about 400 times the Moon's distance, and also by pure chance the Sun's diameter is about 400 times the Moon's diameter. This means by a matter of simple proportion the angular size of the Sun's and the Moon's discs in the sky are almost exactly the same.

Both disc sizes actually vary slightly. This is because the orbit of the Moon around the Earth, and the Earth around the Sun are slightly elliptical and so distances vary a little. This is what gives differing times to the period of totality from usually about two minutes to even occasionally as much as five.

As the Moon's orbit round the Earth is slightly tilted to the plane of the Earth's orbit round the Sun, the Sun, Moon and Earth are generally not quite in line at the time of New Moon[2], so an eclipse cannot happen. But just occasionally when the alignment is exact, we get an eclipse. A total eclipse is seen very rarely from Britain, but if you are willing to travel it is possible to see one most years somewhere in the world. You need plenty of time and money. It is an occupation for the rich retired!

The most striking feature of a total, as distinct from a merely partial, eclipse is that the when the Sun's bright disc – the photosphere – is cut out, the sky goes dark, so you have two or three minutes of night time in the middle of the day. The phenomenon of 'daylight' and the atmosphere was considered in theme 4 'our protective blanket'.

Daylight, which we take so much for granted, is simply sunlight, scattered and blended by the atmosphere. So when it is overcast and maybe we don't actually see the Sun itself for days, we still get the benefit of this filtered sunlight all around us below the cloud level. The sky may not be blue, but we can see where we are going. However at the time of an eclipse the Moon's disc acts like a shutter, literally cutting off all the Sun's rays, well above the atmosphere, so 'daylight' cannot happen and instead the sky is dark. We have the strange experience of two or three minutes of night sky in the middle of the day! During the eclipse we do however see a beautiful white halo around the Moon – the Corona – the Sun's outer atmosphere, which cannot usually be seen as normal daylight is far too strong and obscures it.

*****

About the ninth hour Jesus cried out in a loud voice, "Eloi, Eloi, lama sabachthani?" which means "My God, My God, why have you forsaken me"

Matthew 27:46

There is a very primitive psychological aspect to a total eclipse, as well as its physical splendour. I believe it is to do with the feeling of being **cut-off** from our normal sustaining source of energy. It is actually rather frightening. It almost instantly gets dark and the temperature drops steeply. We know of course with our minds that this artificial state of affairs will only last two or three minutes, and then we can return to our familiar comfort zone of warmth and light. It is almost with a feeling of relief that we see the beautiful 'diamond ring' effect as the intense disc of the Sun again bursts out from behind the Moon, and very quickly it becomes light and warm again.

Cats and dogs hate this surprising and rather sinister phenomenon and are best kept indoors. Bird life becomes totally disorientated for a few minutes as for an instant all their normal 'props' of life are disrupted.

It is sobering to reflect that such a total cut-off of the Earth's regular power supply, if sustained, would quite quickly lead to the end of life on the planet. Life would carry on, using stored energy for a little while but then, rather like a battery running down without being recharged, the Earth would become lifeless. Yes it would continue to obey the laws of motion and orbit the Sun like all the other planets, but it would be a dead world. We have two minutes watching a total eclipse to remind ourselves of our total dependence on the Sun as our source of energy and to experience the mercifully brief but very real sensation of having our ultimate power supply cut off, like a brief taste of what it would really be like.

As a Christian I find the phenomenon of an eclipse helpful in giving me just a tiny glimpse into what it must have been like for Jesus, dying on the cross, when God actually turned his face away from his own son. Jesus was bearing the burden of our sins, and God could not bear to look upon sin. For the first time ever in his earthly life Jesus was actually out of communication with his father in heaven. This must have been the hardest thing of all to bear, frightening, total desolation and utter loneliness, on top of the physical agony of crucifixion itself.

We have seen that for us under a normal overcast sky, we still have light and warmth, thanks to the atmosphere. This is a comforting fact at times when we are going through a 'dry' period when God does not seem to be around and we can't trace his hand in events. Really he *is* still there, like the Sun behind the clouds, continuing to care and provide for us, even though we feel for a time cut-off. But for Jesus dying on the cross the situation was quite different. This was **total eclipse**. It is really impossible for us to comprehend the scale of the solitude experienced by Jesus cut off from his own life-blood, the Heavenly Father with nothing to comfort or sustain him through his ordeal.

*****

1. total eclipse of Sun – when the Moon's disc totally covers the Sun's disc in the sky, thereby cutting off all direct Sunlight.
2. New Moon – the moon is at the zero point in its phase cycle, passing close to the Sun in angle, and hence invisible, as it is daytime.

Conclusion   Drawing the ends together

The Anglo-Australian telescope
3.9 m reflector

Praise the Lord from the Heavens, praise him in the heights.
Praise him, all his angels, praise him all his heavenly host.
Praise him, Sun and Moon, praise him, all you shining stars.
Praise him, you highest heavens, and waters above the skies.
Let them praise the name of the Lord,
for he commanded, and they were created.
He set them in place for ever and ever.
He gave a decree that will never pass away.
Praise the Lord.
Psalm 148:1-6

The twelve themes contained in this book of reflections are my personal set of thoughts about God and the Universe. Whether you agree with them or not is equally a personal matter for you. If you are a non-believer and in particular an atheist, you may find yourself totally rejecting my continued assumption that behind every act or event in the universe there is God, the Creator. Or you may be an agnostic, open minded to the arguments, but still not finding sufficient grounds for belief. You may be a believer in God in a rather abstract way – believing that he exists, set it all up, and is possibly there somewhere behind the scenes 'keeping an eye', but you find it difficult to see how such a great and lofty personage could possibly have any form of actual relationship with his created beings.

People sometimes ask if being a scientist, and particularly an astronomer, helps or hinders a belief in God. This is a good question. I believe the answer is that it all depends where you start from. This sounds a little evasive! It is not meant to be. If you start out at the beginning of a scientific education **without** a faith in God it is very possible that a study of the universe, its vastness and apparent coldness tends to alienate. It seems more and more a lost cause that man could ever possibly be in touch with the Creator, if indeed he actually were to exist, and any form of 'proof' appears impossible. If on the other hand you start out **with** a belief in God, then certainly an increasing awareness of the amazing scale, detail and grandeur of the universe can have the effect of deepening your faith and helping your concept of the nature of the God in which you already believe.

I am absolutely sure you simply cannot prove (and equally cannot disprove) the existence of God through a study of astronomy, or any other branch of science. I asserted in the introduction, science and religious belief are two parallel tracks with complementary, but different, agendas. Christians believe God took the initiative and sent His own son Jesus Christ to Planet Earth, to be a model on which we can base the conduct our lives, and to make the supreme sacrifice of dying for our sins in our place.

This is the only way we can be reconciled with a righteous God who cannot countenance sin, and to enter into a full relationship with him. Science isn't going to help us at all with the step of faith needed to accept this. But once it is accepted, we begin to have a new insight into the nature of God.

We then have a basis for appreciating more fully the wonders of God's creation, and relating to the Living God as a wonderful and caring Heavenly Father, rather than very remotely as the dispassionate 'winder-up of the spring'.

I hope you have found these reflections interesting and that they have given you some food for thought. If the result has been to bring you a little closer to discerning the Loving God behind the universe I would be very pleased.

> *'For I am convinced that neither death nor life, neither angels nor demons, neither the present nor the future, nor any powers, neither height nor depth, nor anything else in all creation will be able to separate us from the love of God that is in Christ Jesus our Lord'*
>
> *Romans 8:38-39*

About the Author

Dr. Frank Flynn read astronomy at University College London, and then went on to do his Ph.D. in astronomical optics at Manchester University.

For most of his working life he has been a teacher of mathematics, and for ten years was headmaster of Mildenhall Upper School. For some years he was chief examiner for GCSE astronomy, and regularly taught astronomy to school students and adult groups. He now works for the Cambridge University Institute of Continuing Education as a part time tutor in astronomy, giving courses in different branches of astronomy to the general public.

Frank Flynn is a committed Christian and passionately believes a scientific view of the universe can be happily harmonised with the Christian faith.

Printed in the United Kingdom
by Lightning Source UK Ltd.
120775UK00002B/235-243